HIDDEN RODS/HIDDEN NUMBERS

Manon P. Charbonneau

Cuisenaire Company of America, Inc. • 10 Bank Street, P.O. Box 5026, White Plains, N.Y., 10602-5026

HIDDEN RODS/HIDDEN NUMBERS
Simple Games of Logic

Manon P. Charbonneau
The Learning Center at Santa Fe

To all readers of this "idea booklet:"
These ideas are dedicated to all the children with whom I have had the privilege of working in the past fourteen years. The creation of HIDDEN RODS is wholly theirs, and I act only as their agent in getting their thoughts out to other teachers and children around the world.

To teachers I say, "GET OUT YOUR BOX OF RODS AND PLAY." Play the games as they are described herein: try the children's own words and ideas, and take their freshness and enthusiasm back to your own classroom.

HIDDEN RODS AND LOGIC GAMES

"Close your eyes. O.K. Now you may open them. I have a rod hidden in my pocket. You may ask me questions about it to help you guess which rod it is, but for right now you may not ask a direct question like 'Is it red?' or 'Is it green?' You may ask questions like 'Is it shorter than yellow?' or 'Is it longer than a train of a white and a light green?' I will answer you directly with 'Yes, my rod is . . . ,' or 'No, my rod is not . . .' And from these clues you should be able to figure out which rod I have in my pocket."

And from there the children are off and running, spilling over with questions, eager to be the first to guess which rod I have hidden. The game was devised for the purpose of finding yet another way to increase young children's awareness of the size and shape and color of Cuisenaire® rods, besides the building and making of designs with the rods which all children, whether they are pre-schoolers or "sophisticated" junior highs, find intriguing. The very simple identification game blossomed into a most fascinating and intellectually profitable game for children.

My practice in the classroom, whether it is a traditional or more open one, has been to introduce these hidden rod games through conversation — no written clues are given at all. I would highly recommend this approach to teachers, though it is not the only way. You must take into consideration the children you are working with, and determine for yourself which is the most appropriate. My favorite approach is, "I have a new game. Who'd like to play?" I generally get five or six "takers." After we've played the game a lot, I try, "I've worked out a real tough hidden rod game. Who'd like to try?" Once I have assembled a little group, I introduce three or four different games, and then ask the children to make up one they'd like to give the others. After a while I can step back to listen and watch as the children take over the game — the situation I strive for in my classroom. I prefer to listen carefully in order to determine the children's pattern of thinking, so that I may better guide their learning into profitable channels for them.

Keep the games simple at first, particularly with the very young. With one five-year-old group I did nothing but "hide" one rod at a time for many, many weeks, until the children themselves clamored, "Hide two rods. Please hide two rods." That was my clue that they were ready for a more complicated set-up. From that point on there was no stopping them, and they were playing the game among themselves with very little or no teaching supervision. One of the most successful settings for the game of hidden rods is a vertically grouped class of fives, sixes, and sevens, where the older pupils can organize and play the game with the younger ones. A tape recorder to record the questions, answers, and conversation is a fantastic aid for the teacher.

I have a very simple reason for keeping the games oral for quite some time: children of all ages need experience talking, expressing themselves creatively and clearly, and they also must learn to get facts and inferences by listening. This is NOT just a question of paying attention — children must learn to take in information through the ear, and encouraging this type of play can help develop this skill. One way to help is to ask after each question is answered, "Which rods could it (they) be?" and "Which rods can we say for sure they aren't?" This kind of questioning helps children sort out

needed information. Oral games also encourage the children's listening to each other, which is as important as their listening to the adult. Make no judgemental comments such as, "That was a good question." Let the children see that some questions and clues are better, that is more useful, than others.

When the children are ready, and only you as their teacher can determine when, begin introducing two or three clues at one time which determine the hidden rod or rods. I continue to actually pick up and hide rods, in my hand, in a pocket, or in any convenient place. *I* personally need the security of having them hidden because I often become so engrossed in listening to the children that I forget what I have hidden.

A game of multiple clues might be: "I have three rods in my hand. The first is $1/2$ of the second, and the second is $2/3$ of the third. Can you tell exactly which rods I have? Or, are there several possibilities? What other kind of a clue would you like? What would help?" Get the children to establish the kinds of clues that help produce a unique solution, for that is the goal of this game: a set of clues that will uniquely determine a set of hidden rods. Along the way there are many, many side roads to be explored, each of which will be briefly discussed below. I believe it takes fun out of playing with the children if you, as teacher, know too much to begin with. So don't be afraid to play with the children and find out with them.

There are many ways the game can go once the children can handle several clues at a time. For me, one of the most effective methods has been with written answers to the children's spoken questions. A sample game might go like this: (I say *might* because everything with children is so unpredictable and should remain so — play with what you get, build on it, and make the most of what they are ready to talk about): Pick your rods, hide them, and tell the children how many blocks you have by writing: $\square$, $\triangle$ That says, "Two blocks!"

The next question from the children is usually something like, "Are the two blocks alike? The written answer can be, $\square \neq \triangle$

The next question might be, "Is the train longer than orange?" $\square + \triangle \not> $ orange

"Is the train longer than yellow?" $\square + \triangle > $ yellow

This establishes that the train is shorter than orange but longer than yellow. Possibilities?

"Is the shortest rod 1/2 the longest?" $\square = $ 1/2 $\triangle$

Is the set "unique?" Which rods can they be? What would be a good next question? Or, what do you think the kids might ask? Anticipating some of their thinking is great fun.

Another approach is to give perhaps three written clues and see how far the children can go before they need another clue. Ask them to establish the next clue and see if their clue fits the rods you have in mind.

With older children, giving repetitive clues is not only great fun, but helps them develop a critical eye and ear. Sometimes, especially with very keenly interested older children, give six to eight, or even ten clues, and ask them to cross out all the extras. In other words, get it down to the fewest clues that describe the answer.

Once interest and proficiency in the game are established, encourage the children to write their own clues on cards or paper. Younger children may need your help. Collections of clues can be shared among the children, maybe among classes in the school, and may be reproduced into a little booklet. I have collected three distinct sets of games. Set one (cards 1-30) was created by five, six, and seven year olds talking into the tape recorder. The children spoke their clues into the recorder, and I wrote them out as they were spoken. Some of the language is spectacular, keenly perceptive, and very intuitive. The second set (cards 31-60) was written by eight, nine, and ten year olds, all of whom were enjoying mathematical symbolism as a really unusual way to express oneself. The third set (cards 61-78) was written by older children, eleven and twelve year olds, who had become fascinated with things such as primes, odds, evens, what happens if you add two consecutive primes, etc. They developed a collection of real "toughies." The cards in this book are samples of each set, samples to help you get started playing the games with pupils and encouraging them to write their own cards. What they create, with your guidance, will be much more meaningful for them than a set of project cards handed to them for them to solve. When recording games, immediately try to go along with a language and a code system that is meaningful to the children. You can gently lead them into eventually using correct mathematical terminology. Talk to the mathematicians or other available resource people for ideas if you wish. But initially, be light-hearted and light-handed with the children. They will have to see for themselves that some sort of standard notation is necessary for universal understanding before they can accept and use it.

Above all, don't be afraid to try the games with children, and don't be afraid to back out when you are no longer needed to lead the games. Learn and play and have fun with the children.

One last idea: just very recently some of my older kids began playing around with rods in a very different way. They invented a game whereby the clues they would toss out were the "value" of the white rod and the "value" for the orange rod, and the "game" is to find a value for each of the other colors that is consistent with the two clues given. There were very many "obvious" ones, such as, if white is 10, orange is 100,

if white is $\frac{1}{10}$, orange is 1,

if white is 12, orange is 120,

where it was quite easy to figure the values for the other colors. Then, one day Scotty came in with several really challenging ones. He had begun to look at rods in ways other than strictly numerical. The whole thing became quite an obsession with many of us, and we would greet each other daily with, "I've got one for you! If white is . . . , then orange is . . ." The fun was in coming up with one that took hours to solve and really twisted the brain cells.

I want to leave you with several that are tough but for which we figured out an answer and several of which no one except the author of that game has worked out satisfactorily. If any of our readers can find a good solution to those, let us know.

If white is 1, orange is 1.
If white is 12, orange is 12.
If white is 8, orange is 8.
If white is 6, orange is 42.

Now have a crack at some of the as yet unsolved ones.

If white is 1, orange is 7.

If white is 1, orange is $\frac{1}{3}$.

If white is 1, orange is 4.

Consistency with the two clues given is the only requirement for a correct solution! There may be many ways of looking at the "problem" which would produce good consistent solutions that are very different from the usual mode of thinking. Remember — rods have attributes other than length! These kids had totally freed themselves from the usual academic restrictions, and were playing around with ideas, and a whole new way of looking at the world. On the last day of school, several of the kids wanted to tell me their "game," but I steadfastly refused to be told as I want some more time to work it out for myself.

Four rods.

All different.

Longest rod shorter than yellow.

Three rods.

All different.

Train is as long as dark green.

Two rods.

Train is as long as brown.

Difference between the two rods is red.

Three rods.

All different.

Train is as long as blue.

Longest rod is shorter than yellow.

Three rods

No rod is longer than light green.

Train equals blue.

5

Three rods.

All different.

Difference between the longest and shortest is red.

Longest is black.

Two rods.

Train equals orange.

Difference between the two rods is zero.

One rod.

It is smaller than light green plus red.

It is larger than light green.

Three rods.

No rod is longer than yellow.

No rods are twins.

Train equals black.

9

Five rods.

Train is longer than purple but shorter than orange.

All the rods are alike.

Two rods.

Train equals red times brown.

The rods are twins.

11

Five rods.

If you make a staircase, the step between each rod is red.

Red is not one of the rods.

One rod.

It is shorter than red plus yellow.

It is longer than brown minus light green.

13

Two rods.

Both rods are longer than green.

Train is longer than orange.

No rod is yellow.

Yellow plus the shorter equals the longer.

Two rods.

Difference between the two rods is white.

Shorter is $\frac{3}{4}$ of the longer.

15

Two rods.

One rod is $\frac{1}{3}$ of dark green.

The other is 3 times as long as the shortest.

Two rods.

Difference between them is light green.

The train is as long as orange plus light green.

Four rods.

Train equals two orange minus a red.

Purple is one of the rods.

Yellow is one of the rods.

There are only two colors.

Three rods.

Train is smaller than purple.

19

Seven rods.

Six of the rods are the same color.

Train is as long as three orange plus
a dark green.

Five rods.

Only two colors.

Train of four smallest rods equals the largest rod.

Largest rod is longer than yellow but smaller than blue.

Four rods.

Train equals brown.

Difference between the longest and shortest is zero.

Difference between the middle two is zero.

Three rods.

All rods are different.

No rod equals four whites.

No rod is longer than yellow.

Train equals orange.

Two rods.

Smallest is $\dfrac{2}{5}$ of the largest.

Largest is not bigger than yellow.

Smallest is bigger than white.

Three rods

Train equals orange plus red.

One of the rods is $\frac{1}{2}$ of the train.

One of the rods is $\frac{1}{3}$ of the train.

Four rods.

All are the same.

Train equals orange plus dark green.

Three rods.

Two rods are twins.

The largest is the same as a train of the two smallest.

Largest is smaller than light green.

27

Two rods.

Two colors.

One rod is $\frac{1}{4}$ of purple.

The train equals $\frac{3}{4}$ of purple.

Five rods.

The train equals dark green.

Four rods.

None are the same.

The train equals three yellow minus purple.

2 rods

No duplicates

w = 1

Longest < 6

Train = 7

Longest − shortest = 1

No duplicates

w = 1

Longest < 6

5 × shortest < 6

5 × longest < 24

Longest − shortest = 2

2 × shortest is not one of the rods

Longest = 3 × shortest

5 rods

No duplicates

w = 1

Longest $<$ 6

Train = 15

4 rods

No duplicates

w = 1

Longest < 6

Longest − shortest = 4

Train = an odd number

One of the middle rods is 2 × shortest

2 rods
No duplicates
w = 1
Longest < 6
Longest − shortest = 4
Train = an even number

35

2 rods
No duplicates
w = 1
Longest < 6
Train = 3
Longest − shortest = 1
Longest + shortest + 7 < 12

3 rods

No duplicates

w = 1

Longest < 6

Train > 10

If you multiply each rod by 3, each product is less than 16

Even number of rods

No duplicates

w = 1

3 × longest < 18 − 5

Train < 16 − 9

Longest < 4

1 rod

w = 1

Train < 3

The rod is odd

Number of rods is even

No duplicates

w = 1

Longest × 3 < 16

Longest − shortest ⩽ 3

Longest + shortest = sum of the other rods

Train is even

5 × each rod gives 10, 15, 20, and another product

w = 1

Longest $\times$ 3 $<$ 16

Longest $-$ shortest $\leqslant$ 3

2 (Longest $-$ shortest) = 6

Shortest + another rod = longest rod

2 $\times$ shortest + 2 = 6

1 pair of twins

Train = 15

4 rods

Longest × longest < 26
Train = 16
Fewer than 5 rods
No more than 2 of any rod
Shortest ÷ 2 leaves a remainder of 1
One rod × 7 = 35
Another rod × 7 = 21

w = 1

Longest × 5 < 10

Number of rods > 5

Shortest × shortest = shortest

2 times each rod, then add them = 24

Each rod has a twin

43

w = 1

Longest × 8 < 48

4 rods

Longest − shortest > 3

2 rods < 3

2 pairs of twins

3 times each rod, then add them = 36

w = 1

Longest × 9 < 49

6 rods

Train = 19

Longest + 2 = 7

Longest + 1 = number of rods

Longest − shortest = one of the rods

Longest − shortest is a twin

3 × longest + 4 × shortest = Train

Even = odds

w = 1

Longest × 8 < 48

5 rods

Longest × 8 > 32

Shortest × 7 > 20

Train of evens = 8

w = 1

Longest × 9 < 46

6 rods

Longest × 8 > 35

2 × shortest + 7 × longest < 38

Train = 18

w = 1

Longest ÷ 5 = 1

Train < 36

Train ⩾ 21

Number of rods = longest + 2

Longest − 1 is a rod

Longest − 2 is a rod

Longest − 3 is a rod

Longest − 4 is a rod

2 × one twin is another twin

$w = 1$

longest $\times$ 5 $<$ 26

Longest $\times$ 5 $=$ number of rods

Longest $\times$ 4 $<$ 20

Shortest $\times$ 7 $<$ 27

Number of rods $=$ 10

w = 1

Longest × 13 ≤ 65

Longest × 17 > 70

Number of rods = 6

Train = 18

Rods are all odd

(Longest − shortest) − 1 = one of the rods

There are not equal numbers of each rod.

w = 1

No duplicates

Train = 40

Number of rods = 5

w = 1

No duplicates

Train = 15

Number of rods > 2

Longest < 6

w = 1

No duplicates

Train = 15

Number of rods > 2

Number of rods < 5

Longest − shortest = 6

Longest × 5 ≤ 39

Two times one of the rods = 6

Add any two rods (not longest or shortest)
$\qquad$ **and you get longest**

w = 1

No duplicates

Train = 37

Number of rods = 7

Shortest × 7 > 7

Longest × 7 > 28

Shortest < 3

w = 1

No duplicates

Number of rods = 4

Train = 18

Longest = one of the rods × 3

Shortest × 7 < Longest

w = 1

No duplicates

Number of rods = 14 − 10

Train = 18 + 7 − 5

$$\frac{\text{Longest}}{2} = \text{1 rod}$$

Shortest × 2 = one rod

w = 1

No duplicates

Number of rods = 5

Train = 25

$21 \div 3$ = one rod

$24 \div 8$ = one rod

$117 \div 117$ = **Shortest**

Longest $\times 5 \neq 50$

No evens

w = 1

No duplicates

Number of rods = 5

Train = 22

Longest − shortest = 8

Longest − 2 = one rod

w = 1

No duplicates

Number of rods = 4

Train = 13

Shortest × 2 = one rod

Shortest + 2 = one rod

Longest − 4 = one rod

w = 1

Number of rods = 3

Train = 30 − 6

Shortest is even

All others are odd

I am thinking of

A) 3 numbers

B) their sum is 24

C) the numbers are all even

D) the largest number < 14

Stop here until you think you have
found the numbers.

(continued on 62)

61

Have you found a unique set
of numbers to fit the clues?
Is there more than one set that
will fit? If so, how many sets
can you find? List them all.
Do you need another clue to find
my unique set? Here it is.

E) The smallest number is 2
How can you determine the unique set?

What is it?

Here are clues for a set of Hidden Numbers:

    A) the numbers are all prime

    B) their sum is 50

    C) the largest number < 30

    D) there are 3 numbers

Are there enough clues to determine
    a unique set?

Find these hidden numbers:
- A) the numbers are all odd
- B) their sum is 16
- C) the largest number is 7
- D) there are 4 numbers

Do you need another clue
to find a unique set?

If so, what kind?

New game. The clues are:

    A) there are 4 numbers

    B) their sum is 10

    C) largest minus the smallest

        equals 2

    D) the product of the smallest
       times the largest is odd

Only two clues for this one!

A) there are 5 numbers

B) their sum is 6

Here are 2 clues:

    A) there are 3 numbers

    B) their sum is 21

What are the possibilities?

Here's another clue:

    C) the average of the 3 numbers

       is 7

Now can you tell what the 3
numbers are? If not, why not?
If you need another clue, try the card 68.

D) the largest minus the

smallest equals 9

What are the numbers?

Clues:

> A) there are four numbers
> B) there are no duplicates
> C) their sum is 10

What if I changed the last clue to read:

> C) their sum is 11?

Is there still a unique set, or
are there other possibilities?

Try something different. Here are
clues A and C.
You can supply clue "B".

   A) 3 numbers

   B)

   C) Largest − smallest = 0

What kind of clue would you
write for B?
Do you think your friend will think of
the same numbers?

Clues:

    A) 1 number

    B) it is greater than or equal
              to 6

    C) the number squared is
              less than 100

    D) the number + 5 is less
              than 14

    E) it is even

Do you know the number?

Clues:

A) 1 number

B) it is less than 45

C) it is a multiple of 6

D) twice the number is greater than 60

E) it is not a multiple of 4

The number is ☐ .

72

Clues:

A) 1 number

B) it is prime

C) the number + 1 is an odd number

73

Clues:

A) 1 number

B) the number + 1 is a
multiple of 4

C) 3 times the number is
less than 150

D) the number divided by
2 is greater than or
equal to 15

E) the number is a multiple of 3

Clues:

A) it is a fraction

B) the denominator is four
more than the numerator

C) the numerator is less
than 6

D) The denominator is greater
than 6

E) the fraction is greater
than $\frac{1}{2}$

75

Clues:

A) the number is a fraction

B) when you add the numerator
to the denominator the
sum is 21

C) the fraction is greater
than $\frac{1}{3}$ but less than $\frac{1}{2}$

There is 1 number. When you write
it in the box, each of these becomes
a true statement:

$$\Box \geqslant 6$$

$$\Box \times \Box < 100$$

$$\Box + 5 < 14$$

$$\Box \text{ is an odd number}$$

What is the number?

There is 1 number.

When you write it in the box, each

of these becomes a true statement:

$\square$ + 2 is a multiple of 5

3 × $\square$ < 150

$\square$ ÷ 2 > 15

$\square$ is a multiple of 3

What is the number?

ANSWERS

1. w, r, g, p
2. w + r + g
3. g + y
4. r + g + p
5. 3g
6. y, d, k
7. 2y
8. p
9. w + r + p
10. 5w
11. 2n
12. w, g, y, k, e
13. d
14. p + e
15. g, p
16. r, d
17. y + n
18. 2p + 2y
19. 3w
20. 6y + d
21. 4r, n
22. 4r
23. r + g + y
24. r, y
25. r + p + d
26. 4p
27. 2w, p
28. w+r
29. 4w + r
30. w + r + g + y

31. g + p
32. w, g
33. w + r + g + p + y
34. w + r + g + y
35. w + y
36. W + r
37. r + p + y or g + p + y
38. w + r or w + g or r + g
39. w
40. w + r + g + p or r+g+p+y
41. r + g + 2y
42. g + 2p + y
43. 12w
44. 2w, 2y
45. 2w + 3p + y or w+r+g+2p+y
46. 2p + 3y
47. 3w + 3y
48. w + 2r + g + 2p + y
49. Any set of 10 rods, whites and reds only, with at least one red.
50. w + 4g + y
51. d + k + n + e + o
52. w + r + g + p + y
53. w + g + p + k
54. r + g + p + y + d + n + e
55. w + g + y + e
56. r + p + d + n
57. w + g + y + k + e
58. w + r + g + k + e
59. w + r + g + k

60. d + 2e
61. 2 + 10 + 12
    4 + 12 + 8
    4 + 10 + 10
    6 + 12 + 6
    6 + 10 + 8
    8 + 8 + 8
62. 2 + 10 + 12
63. Yes; 2, 19, 29
64. Need a clue about triples or no duplicates.
65. 1, 3, 3, 3
66. 1, 1, 1, 1, 2
67. No unique set.
68. 2, 8, 11, or 3, 6, 12 or 4, 4, 13
69. 1, 2, 3, 4 — sum is 10
    1, 2, 3, 5 — sum is 11
70. The three numbers must be the same, but can be any number.
71. 8 or 6
72. 42
73. 2
74. 39
75. $\frac{5}{9}$
76. $\frac{6}{15}$
77. 7
78. 33